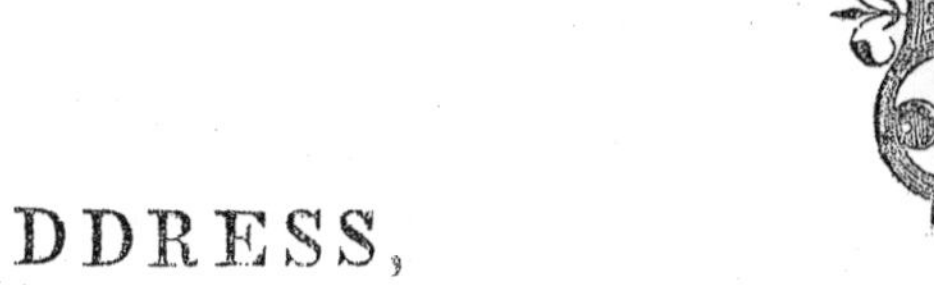

ADDRESS,

DELIVERED OCT. 2D, 1851, AT THE TENTH ANNUAL EXHIBITION

OF THE

QUEENS COUNTY AGRICULTURAL SOCIETY,

AT JAMAICA, L. I.

BY THE HON. JOHN A. DIX.

PRINTED AT THE OFFICE OF
THE HEMPSTEAD INQUIRER,
HEMPSTEAD, L. I.
1852.

ADDRESS,

DELIVERED OCTOBER 2D, 1851,

AT THE

TENTH ANNUAL EXHIBITION.

OF THE

QUEENS COUNTY

AGRICULTURAL SOCIETY,

AT JAMAICA, L. I.,

BY THE HON. JOHN A. DIX,

PRINTED AT THE OFFICE OF
THE HEMPSTEAD INQUIRER.
HEMPSTEAD, L. I.
1852.

HEMPSTEAD, L. I., October 2d, 1851.

HON. JNO. A. DIX,

Dear Sir:

At a meeting of the Managers of the Queens County Agricultural Society, held this day, a resolution was adopted tendering the thanks of the Society to you for the able and eloquent address which you have just delivered, and requesting you to furnish a copy for publication.

I have great pleasure in communicating this request, and indulge the hope you will comply, at your earliest convenience.

With great respect, I am, sir,

Your obedient servant,

JNO. HAROLD, Secretary, etc.

PORT CHESTER, Oct. 7th, 1851,

Dear Sir:

I have the pleasure of acknowledging your favor of the 2d inst., communicating a resolution adopted by the Managers of the Queens County Agricultural Society, requesting me to furnish for publication a copy of my address delivered before the Society on that day.

I am about to leave home, to be absent for some weeks. Should it be in my power after my return, to write out my address, which was delivered from notes, I will do so; and in the meantime,

I remain,

Very respectfully,

Your obedient servant,

JOHN A. DIX.

JOHN HAROLD, ESQ.

ADDRESS.

Mr. President, and Gentlemen:

If I had been governed by a consideration of my qualifications for addressing you on the subject of Agriculture, I should certainly not have ventured to accept the invitation of your Committee. I have really very little knowledge of practical husbandry. My occupation until quite recently has not been such as to fit me for making any useful suggestion to you, the farmers of Long Island, known, as you are, throughout the State for your familiar acquaintance with all that concerns a successful cultivation of the soil. But the interest I take in the subject has overruled all other considerations, though in appearing before you, I am compelled to throw myself on your indulgence and to call your attention to topics more remotely connected with agricultural life and occupations than those, which are usually discussed on occasions like this.—Indeed, gentlemen, I feel that I should justly incur the imputation of presumption if I were to undertake to advise you as to the rotation of crops, the raising of domestic animals, the preparation of manures or other subjects of a kindred character—subjects, on which you are much better informed than myself. It is under the influence of this conviction that I turn to other topics, collateral to these, and I trust intimately interwoven with the lasting interests of an Agricultural community.

Before I proceed, to the discussion of these topics, let me call your attention to some local considerations, which concern you as residents of Long Island. I believe I hazard nothing in saying that few other Districts in the State possess higher advantages. As an Agricultural District alone, these advantages are inappreciable. Your County lies upon the very confines of a City destined to become one of the most populous, in the world, and increasing with rapidity altogether without a parallel. A circle with its centre at Union Square in New York, and with a radious of two miles in extent, will embrace a population of seven hundred and fifty thousand souls. The annual growth of this immense aggregation of people will far outstrip the ability of surrounding districts to supply its wants. Even now it is, by means of railways, reaching into remote Counties and States and drawing forth their agricultural surpluses for consumption. Your proximity to this extended and unfailing market gives you great advantage in the competition. The Sound and the East River afford you a sheltered communication by water. A Rail Road divides your County in nearly a central course. By one or the other of these channels of intercourse any farmer in this County may reach the City in a few hours with the surplus products of his labor, and always with the certainty of finding a market for all which he can spare from the consumption of his household.

Considered in reference to fertility of soil, your County will bear a favorable comparison with others in the southern portion of the State. From its characteristic qualities it is doubtless destined to be devoted almost exclusively to tillage. There may be exceptions in the northern portion of the County, but I believe I am justified in saying that nature indicates its superior fitness for the cultivation of Grains, Fruits and garden products, and that accidental circumstances confirm this application of your agricultural labor.

I remember often to have heard it remarked in former years that the profits of farming on Long Island were absorbed by the expenses of manuring. But I have become convinced by personal observation that the productiveness of your agriculture, labor and manures being both taken into the account, will bear a fair comparison with that of other districts in this part of the State. If much of your soil is light and thin, it is at the same time genial and warm; and the rapidity, with which vegetation matures, gives you an advantage in the market, over districts, in which nature is more sluggish in her operations, and tardy in her returns. The variety, which your soils possess,—from the deep rich loams of your necks to the light sandy formations on the South Bay, gives you the power of a diversified production, which is eminently desirable in a district destined to contribute largely to the consumption of a City, demanding from its magnitnde, a varied supply.

Gentlemen, your County has other advantages, which ought not to be overlooked. Placed as it is, between the waters of the Sound and the

Atlantic Ocean, it is alike exempt from the extremes of winter and of summer, you are neither pinched by excessive cold nor overpowered by enervating heat. I believe I may truly say that there is no region in this hemisphere, which possesses in a higher degree the advantages of health and personal comfort, or which admits of a more uninterrupted application of the physical powers. These are all elements of prosperity, and they should be causes of devout thankfulness to the Sovereign Ruler of the Universe for having so bountifully endowed you with the capacity of combining individual enjoyment with high social welfare.

Though you have some sources of prosperity peculiar to yourselves, arising out of geographical position and internal advantages, it is gratifying to know that most other portions of the State enjoy like facilities, though not all to the same extent, for improving their condition. Agriculture, under a variety of forms will, in all probability, be for a long period of time the ruling interest of this State. The modes of its application may be to some extent varied as the western wilderness is filled up, and the thousands of emigrants who come in from the Old World and the millions of money they bring with them enter into the great field of production. It is not easy to foresee what may be all the results of these accessions of capital and labor ; but I believe they cannot be otherwise than beneficial. I think we may fairly calculate that the standard of Agricultural prosperity will be continually advanced—that the productive industry of the country will from time to time put on new phases, but that under any of its modifications it must of necessity contribute to the advantage of our Agriculture.

Under these circumstances, what are the leading objects to which the attention of the Agricultural community should be turned, with a view to the improvement of their condition ?

First of all is to be ranked the education of children. I have no reference now to the higher aims of intellectual culture—those, which concern the ability of the individual to discharge with intelligence his duties as a citizen, and as an integral part of the political power, by which the machinery of government is directed and kept in motion, or those which concern the moral improvement incident to the cultivation of the mental faculties. Nor is it necessary that I should enter into these considerations. We all know that without an intelligent understanding of the nature of our government, some knowledge of its foreign concerns and its true domestic policy, we cannot do justice to those to whom the political power of the country is entrusted, when they are right, or correct their misconduct when they are in error. So in respect to the influence of education upon the moral condition of a community, I think it cannot be doubted, whatever deductions to the contrary may be drawn from the criminal statistics of particular countries, that crime decreases as knowledge is diffused. Offenses of certain classes—those which are conceived and executed in fraud – may increase as the intellect is sharpened by culture. But those which are perpetrated by force, and in general, those of great atrocity, are undoubtedly diminished in frequency as the standard of popular intelligence is advanced. But I do not allude to influences or objects like those. I hold education to be one of the highest attainments of an Agricultural community as an element of improvement in rural economy. The labor which is most intelligently directed and applied, is always the most profitable. An enlightened farmer will always obtain more from an application of equal means, th n an ignorant one. I do not mean to say that an uneducated man is always ignorant. Knowledge is to be derived from the observation of surrounding objects as well as from the study of books, and a man may become wise in a particular vocation by observation alone. But he would become much wiser if his mind were disciplined by the training of intellectual application. Without this advantage, however high the spirit of inquiry, by which he may be moved, however bountiful his mental endowments, his observations will be less critical and his deductions more tardy and inexact. Education not only leads us to an acquaintance with books, which are the fruits of the observations and reflections, of other men, but it gives directness and force to our own, through the influence of intellectual discipline. This is, indeed, one of its highest offices: it is the great instrument by which obstacles in the path of improvement are overthrown and removed.

The condition of the schools in this immediate neighborhood, is then, an object of leading interest to every farmer. It is of the utmost importanse that the teacher should be capable, honest "and apt to teach." A good teacher is, if possible, of more value than good books. The latter in bad hands, are comparatively worthless. To command good teachers, liberal wages must be paid. This has been the great defect in the district schools. There is no reason under the new system, why it should exist any longer. The expenses of the schools, after expending the money contributed by the public funds are paid almost exclusively by taxes upon property. I know there is a serious difference of opinion in respect to the expediency of such a provision; and I believe the law passed last winter may be regarded as a compromise between those entertaining conflicting views on the subject. I remember when I was Superintendent of the Common Schools of this State, to have come to the conclusion, and so reported to the Legislature on more than one occasion, that Free Schools, as they are called, though it would be more proper to call them schools in which the expenses are wholly paid by public funds and property, (for our schools were free under the old system,) were indispensable to cities and large villages, and that schools in which the patrons paid a small part of the teacher's wages were preferable as a general rule in purely Agricultural districts. But any system of which the tendency is to elevate the standard of instruction in the Common Schools, is the best; and certainly the school tax is the one which above all others we ought to pay with most pleasure. In proportion as the condition of the district schools is improved, the necessity for private schools of an elementary character is obviated. For this reason it is of the first importance that the district schools should be liberally supported. Let this be done, let them have teachers who are well qualified to instruct and to maintain order, and private schools will cease to have any advantage over them and become wholly useless in Agricultural districts. I consider such a result eminently to be desired; for where private schools abound, it is nearly an infallible indication that the condition of the district schools is depressed. It should be regarded, then, as one of the first duties of the farming community to cherish the district schools, to elevate their character, and to consider expenditures upon them as the most productive of all investments.

Next to a preparation of his children for the active duties of life, the farmer will naturally regard with most interest the proper use of the soil he cultivates, so as to ensure the greatest practicable productiveness. I trust I shall not be considered as sinning against the lights of science, when I say that this must in nearly every case be determined by individual observation and experience. The infinite diversity which is found in the chemical composition of soils, even in contigiuous districts and farms, makes the choice of the fertilizing agents best adapted to them, a question to be decided mainly by experiment. In almost every great district, the geological formation of which has a distinctive character, an analysis of the principal soils will furnish valuable hints. In this respect the geological survey of the State has rendered an important service to our Agriculture, by collecting and analyzing the soils of different counties and towns. But these results can rarely do more in any particular case than to afford suggestions of a general character. For instance, in districts in which the soil derives its distinguishing characteristics from the decomposition of lime stone, the analysis will show, as might have been expected, that any manure in which lime is the principal ingredient would be superfluous, if not positively prejudicial. So where the soil depends, upon the debris of granitic rocks or slate, or some of the later formations, for its mineral agents, a general idea of the particular manures best calculated to render it productive, may be obtained from a knowledge of this fact confirmed by chemical analysis. But, as I have already said, only valuable suggestions can be obtained from the results of this analysis. In a district of country varied in its surface as ours is, with alternating valleys and hills, with cultivated fields running up to the summits of mountains and down to the level of the sea, nearly every field may require a specific treatment, and this can only be taught by experiment, for very few farmers have the means of subjecting their farms to a detailed analysis by a practical chemist. In like manner the quantity of land which may be most advantageously devoted to grazing or tillage, the products which from their certainty, or the stability of the prices they bear in market, are likely to yield the greatest average profit, the number of domestic animals, which ought to be sustained in reference to the convenience of the farmer and the proper management of the farm—these are all problems, which, from the necessity of the case, every farmer must ultimately decide for himself, and after a few years of observation he will decide them more intelligently than any one can do it for him.

I have already said, gentlemen, that in these matters I have very little practical knowledge, and I am admonished by a consciousness of my deficiency, in this respect, to abstain from all advice in respect to them.

But there are some points in the economy of rural life, on which I venture to throw out a few suggestions. If the positions I have stated in respect to the future condition of this State as an Agricultural district are well founded, the farmers of New York may fairly calculate, with a diligent application of their labor, on a more than ordinary prosperity. By far the greater portion may reasonably expect a regular surplus of income after meeting the expenses of their domestic establishments. How shall this surplus be applied? in enlarging their farms, or in rendering the land they already occupy more productive by a more liberal application of labor and of fertilizing agents? I believe experience has satisfactorily proved that our farms are, as a general rule, too large already. This was naturally to be expected in the early settlement of the country when land was cheap. But as prices have advanced, the tendency has been to curtail the dimensions of our farms. Still, the evil continues to exist to some extent. There is no doubt that the same quantity of labor expended upon a diminished surface, would be, in most cases, more profitable. This observation is particularly true in respect to farms devoted chiefly to tillage. Very large farms, in nine cases out of ten, will exhibit some traces of slovenly cultivation—a field here and there overrun with weeds, or some other indication that the surface is too extended for the minute observation of the superintending head. He who has more acres than he can conveniently cultivate, had better sell them, and devote the interest of the money to the more thorough cultivation of what he retains. By so doing, his comfort and his pecuniary interest will alike be promoted.

Next to the proper cultivation of the farm, should rank the neatness and order of the farmer's residence, the farm house. It should be tasteful, comfortable, and if possible, of durable materials. The best material, undoubtedly, is the one in which your county does not abound—stone. But brick, of which Queens is a large manufacturer, is a good substitute for it. I am aware that there is a strong prejudice against stone dwellings, from the idea that they are damp; but I believe the prejudice to be wholly unfounded. A stone house properly constructed is as warm in winter, as cool in summer, and as dry at all seasons as any other. In many countries in Europe with climates quite as moist as ours, the dwellings are almost wholly of stone or brick, and they are perfectly free from dampness. They are made so by detaching the plastering from the inner surface of the wall. When a room is plastered directly upon the wall, moisture will show itself in particular states of the atmosphere, and sometimes to such an extent as to run down in streams to the floor. This is not, as is sometimes supposed, because the moisture passes through the body of the wall, but because the wall being a ready conductor of heat, the temperature of the inner surface conforms almost immediately to that of the outer, and in sudden changes from heat to cold, the atmosphere of the room is cooled so rapidly as to deposit on the plastering the moisture it held in solution. This effect is readily obviated by interposing a body of atmospheric air, (no matter how thin,) between the plastering and the wall. It is only necessary that the one shall not be in contact with the other. The most effectual mode is to fer with studs or scantling; and with this precaution. all apprehension of dampness may be discarded.

But whatever material the farmer may choose for the construction of his dwelling, it should be tasteful in design. It costs no more to put materials into a graceful form than it does to throw them together without architectural taste or propriety. The dwelling should be simple, substantial, and unpretending—a type of the independence and the unobtrusive habits of the occupant. Comfort, not show, should be the leading object in the construction of the American farm house. Our government is simple in its structure; the same character of simplicity runs through every department of our social and political organization; and it is eminently appropriate and desirable that the Agricultural community should preserve in their dwellings their rural improvements and their domestic habits, the distinguishing characteristics of their institutions, and set up if possible, an impassable barrier against the growing waste and extravagance of the great towns.

There is no one thing in the way of domestic improvement, in which we have made so rapid an advance during the last twenty years, as in architecture. Happily the era of Grecian pediments and colonnades for private dwellings, has passed away; and it is to be hoped to return no more. The prevailing forms of Italian, Gothic, Norman and English cottage architecture are far more appropriate for dwellings in external taste, and better adapted to the comfort of the inmates. Of these forms the most simple are the most suitable for farm houses. There is a little danger that they may run too much into ornament for good taste. All excess of ornamental work is a waste of money, and what is worse, it impairs the true effect of the edifice, of which it is a part.

A still greater error is to build on a more extensive scale than the means of the proprietor warrant. This is perhaps not so much the error of the country as of the town. I doubt whether in any channel of expenditure there has been so much extravagance as in this, and nothing can be more unwise. We have fortunately no system of entails or rule of primogeniture to maintain overgrown establishments. Fortunes dwindle away with us by the regular operation of law, as rapidly as they are accumulated. I do not remember a large private establishment in the city of New York, twenty-five years old, which has not passed out of the family of him who erected it, and this, not so much on account of the changes which have taken place in the currents of fashion and business, as because no one of the children could afford to maintain it. In building, the proprietor ought always to ask himself whether any one of his children, with the share of the estate which will be likely to fall to him, will be able to support the establishment, and he should be governed by the answer he can give to this enquiry in determining the dimensions of his dwelling. If he does not, the chances are ten to one that his descendants, before a single generation has passed away, will be compelled to quit the paternal mansion, and do violence to all the endearing associations which, from our very nature, connect themselves with the natal roof and the family fireside. For this reason, it is difficult to see a magnificent mansion rearing itself upon the hill-side, or the bank of a river, overlooking more humble structures and seeming to exercise a manorial supervision over them, without thinking that the unconscious proprietor is in all probability building for strangers, and not for his own children. For the same reason it is always more grateful to contemplate the quiet farm-house, substantial and unobtrusive, standing amid cultivated fields or the tasteful cottage nestling in the valley with its rustic improvements about it.

When I speak of the danger of running into excess of ornament, I do not mean that farm-houses should be without embellishment. Far from it. They ought to be highly ornamented; they should be surrounded with the beautiful and graceful in nature: the vine, the flowering shrub, and such other plants as will bear the rigor of our winters. These are the true ornaments for rural dwellings. They are far more appropriate and tasteful than the most elaborate carvings in wood or stone; and nature offers them freely to all who will take pains once a year to bestow on them a few hours of attention. It is in these appendages to rural dwellings that the great charm of the country in England consists. English farm-houses and cottages are not often, I may say very rarely, faultless structures, when tested by a strict application of architectural rules. Nay, they are often ungraceful in design and rude in execution; but with the ivy spreading itself over the gable or covering up the porch, and the woodbine climbing up the casement and enveloping it in foliage, they acquire a beauty and a grace, which no work of man's hand can equal.

Such as these I should wish our rural habitations to be. They should be embellished, not so much by the hands of the architect, as by the taste and care of the occupants. And I trust the fairer portion of my audience will pardon me for reminding them that this is their peculiar province. Let me, Ladies, address myself to you for a single moment as presiding over the household and the family dwelling. Let it be externally a type of the neatness and order which reign within. Ornament it with the vines, plants and flower-bearing shrubs, which are suited to our climate. These require little attention, and many of them carry their foliage and verdure far beyond the season, when most others decay. Flowers, which require to be housed in winter, demand too much care, and as a general rule, they are in the open air ephemeral in their bloom. The hardier plants, those which come out early and bear their foliage late, are preferable for the decoration of the family dwelling. It is not easy to conceive with how little expenditure of time the most gratifying results may be obtained. A gravelled walk from the entrance gate to the porch, running through a lawn of well cropped grass, with here and there a lilac, an althea or a seringer—a vine trained upon a frame (no matter how rough, for the foliage will cover it) will change the coldest prospect into one of warmth, and beauty, and grace.

Nor is it to the taste alone that these rural embellishments address themselves; they tend to elevate and refine the moral feelings, and to make us better men. It seems difficult to connect with the homestead the sacred feelings which belong to it, when all around it is bare and cold. But when clothed in rural beauty by kindred hands, the sentiment of home is exalted, and those who have thus embellished it, are presented to our minds and hearts under new and more endearing aspects.

I say to you then, Ladies, embellish your dwellings with the beautful in nature; surround them with verdure and foliage, and give them the highest possible attraction to the eye and the taste. The leading impulses by which men are governed, are constantly drawing them out into the world. Ambition, the desire of accumulation, the necessary business of life, are perpetually calling them away from home. Let home, then, be made so attractive in its external as well as its internal aspect, that it shall always be left with regret and regained with eagerness, as the most grateful refuge from the active duties of life. Under these circumstances the minutest work of your hands will have its value. The vine your have trained, the shrub you have planted will possess an interest in the sight of those who are dear to you, which the most elaborate ornament wrought by the hands of the carver can never attain.

I know no works so well calculated to inspire a love for rural beauty, and to teach us how to create it, as those of Mr. Downing of Newbnrgh. I allude particularly to his works on Landscape Gardening and rural Architecture. They should be in the hands of every farmer's wife and daughters. Indeed, there is no one in any condition of life, who may not be improved by their exquisite taste and sound deductions. They have the great merit of teaching the purest morality without seeming to teach it. I do not know that the intellect can ever be educated too much; but I believe it to be quite possible that the moral feelings may be educated too little; and in this respect Mr. Downing's books supply a lamentable deficiency in all that concerns the embellishment of rural life and the beauties of rustic scenery.

Gentlemen, there is no country on the face of the globe which is susceptible in a higher degree of embellishment than ours,—none, which nature has more highly embellished—though it is painful to admit that we have destroyed much which if left where she planted it, would have constituted in itself one of the first elements of rural beauty. I allude to the nearly universal destruction of our noble forests in immediate contiguity with our dwellings. When preserved it is generally for fuel, at a distance from our residences. This error may readily be repaired by planting.

Five, six, or seven years are sufficient to provide a house with shade. I do not think a tree should ever be planted within fifty feet of a dwelling, especially, if it be of large growth. When it attains so large a size as to spread its branches over the dwelling, they tend to perpetuate dampness, and when in contact with the open windows of sleeping apartments in summer nights, the carbonic acid gas they evolve, is often a source of annoyance to sleepers, and sometimes deleterious to the health. A grove five or six rods off, and in such a direction as to shield the house from the bleak winds to which it is most exposed, should be regarded as an indispensible appendage to every farm house.

I think this grove ought not to consist altogether of mere shade trees. Economy and comfort both dictate that it should yield fruit as well as shade; and in sections of the country where forests are disappearing, it is desirable when a tree blows down, or exhibits signs of decay, that the wood may be of such a nature that it may be converted to some practical use. I know few trees more beautiful than the hickory and the chestnut—one is an excellent fire-wood, and the other is valuable for many farm purposes. They have an advantage, too, over most others in their long tap roots, which penetrate deep into the earth, and draw out its secret moisture, so that in the very dryest season their foliage is fresh and green, while that of trees whose roots are superficial, is faded and perishing. Such a grove, beside the comfort it affords to a family in hot weather, will, if properly treated, kept free from undergrowth and seeded down, pay the interest on the value of the land in what it produces. The nuts are valuable for home consumption, or the market: and if the leaves are raked up in the fall and thrown into the barn-yard, to be trampled to pieces by the cattle during the winter, they will add to the manure heap in the spring, and the grove will yield a good crop of grass, though certainly not equal to that of an open meadow. It is only where the leaves are allowed to accumulate in groves, covering the whole surface of the earth, with a thick deposite, which the delicate vegetation can not penetrate, that the grass withers and dies.

There are several other nut-bearing trees which afford excellent shade, and are beautiful in their forms and in their foliage. Of these the black walnut and the butternut, or the white walnut are among the most valuable and graceful.

Let me not be understood as proscribing mere shade-trees. That is not my intention. There are many of great value, which should be cultivated and preserved. There are the sugar maple and the locust. There are the elm and the oak, both trees of noble growth, but they are seen to best advantage standing alone; aud when they are scattered over an extended landscape, and have attained their full size, they have a fitness as emblems of the majestic stature and growth of our country.

To a country embellished and cultivated as ours may and I trust will be, we, and those who are to come after us, will cling with new tenacity as time advances and developes in it new capacities for improvement. It is to the country all men turn at last for a refuge from their worldly cares. Born of the earth, our instincts lead us back, as life draws to a close, to scenes, in which nature presents herself in her most attractive attire, and under her serenest aspects. No man could endure to pass through his last great trial on earth, amid the confusion and uproar of a large commercial city, unless his native instincts had all been obliterated by a whole life of habits at war with them. Statesmen, warriors, men who leave the impress of their character and actions upon the age in which they live, when their public labors are ended, seek the quietude of the country with that instinctive love for its beauties, which seems to be one of the inseparable companions of intellectual greatness. Washington, when his great mission was fulfilled, retired to the shades of Mount Vernon; the elder Adams, at the close of his administration, to Quincy; Jefferson to Monticello; Madison to Montpelier, and Jackson to the Hermitage. Every one of their successors in the administration of the government, has found rest from his public toils in the quietude of rural life.

Connecting themselves, as the best and truest impulses of our nature do, with country life, every one should endeavor to do something to embellish it. The first wish of every man is to live in the memory of his descendants. There is something exceedingly unsatisfactory, not to say revolting, in the thought that they are only to be reminded of us by the bare head-stone, lost among a hundred others in the common burial place, telling in a few words of set phrase, when we entered on the theatre of life and when we made our exit from it. A grove judiciously planted—a few trees set out around the family dwelling, will be far better memorials of us with our children. The after generations who sit down under the refreshing shade, will take pleasure in pronouncing the ancestral names, or in acknowledging the parental providence to which their enjoyment is to be referred.

He who builds a farm house, has a still better opportunity of perpetuating himself in the remembrance of his children. Its tastefulness, its fitness for the purpose of its construction, its internal arrangement for comfort and health, will be to the descendants so many mementos of the ancestor, in whose good judgment they originated. A well built and well arranged dwelling is the best mausoleum a man can erect to himself. He may thus perpetuate his taste or his true conception of domestic order and comfort, and leave to his children a memorial of himself, which shall enter into the business of their lives and mingle itself with their daily occupations and enjoyments. In this manner we may be truly said to live with our descendants.

He who can add to the attraction of Agricultural life by exhibiting it under more interesting aspects, will render the country an inestimable service. Most other pursuits present greater allurement to the eyes of the world, in the higher rewards they sometimes afford, and in the broader fields they open for public distinction. But it is with professions as with every thing else; those which promise the highest rewards, are exposed to the greatest hazards. While one lawyer amasses a fortune, a hundred procure a bare subsistence, and in our cities many go down to early graves, with constitutions broken by sedentary habits. While ten merchants become rich, ninety become bankrupt, as our commercial statistics show. Agriculture is the safest of all pursuits, the most independent, the most healthful, and the least laborious. It is true, it affords no examples of sudden elevation from poverty to wealth; it opens no field to speculation or perilous adventure, But its rewards are certain, and with industry and perseverance, it is sure to yield the competency, which, the wise man has said, is better than great riches.

More than three quarters of the entire population of the United States are said to be engaged in the pursuits of Agriculture; and it will be well for the country, if this proportion can be maintained. With our western boundary extended to the Pacific, and with the new stimulants which the riches of California have offered to commercial adventure, it would not be surprising if there should be some diversion of capital and labor from the field of Agriculture. But this diminution of the effective force of the Agricultural classes cannot be of long duration. The immense forests which are yet to be subdued, beyond the great Lakes, will attract new laborers, and soon repair what Agriculture may temporarily lose. It is not probable that for a long course of years there will be a sensible diminution of the numerical proportion which the Agricultural classes bear to all others. I consider the perpetuity of our institutions as depending in an eminent degree upon the great preponderance of those who are devoted to the cultivation of the earth. I should regard it as exceedingly unfortunate if either the commercial or manufacturing classes, were to acquire a superiority in numbers over the agricultural. They belong to the same great industrious family, but they belong to it in the relation of subordinates. Bound to each other as they are in the general economy of the system by indissoluble ties, it is nevertheless in the steadiness, the quietude, the patient industry of Agricultural life and the habits of sobriety it creates, that our greatest security consists. Pursuits, which bring men permanently together in large masses, have their peculiar dangers.—No man can compare the tranquility which reigns throughout the country, with the constant excitement which agitates the great cities, without feeling that it is to the former we must look for safety against any internal shock, which shall threaten the social edifice with destruction.

When we contemplate the scenes which some of our commercial towns have enacted within the last ten years, classes and associations warring upon each other, lives sacrificed in attacking property or defending it, multitudes wild with excitement, harnessing themselves, as it were, to the car of youth or beauty, or artistic talent, we cannot but feel that it is to the calmer and purer atmosphere of the country, where men think first and act afterwards, that we must look for the governing power, by which the motion of the social machine is to be regulated.

Above all things, gentlemen, let us maintain by every means in our power, the dignity of labor, and especially the labor which fertilizes and embellishes the earth. It is the true vocation of our race. It is one of the merciful designs of Providence that in carrying out the divine sentence, our prosperity and our happiness should be alike promoted.—Let every man, who lives in the country, till the earth. No matter how limited the surface on which his toil is expended. If it be but a garden or a grass plot, let him cultivate it. His health, his feelings, and his character will all be improved.

And finally, let us remember that in maintaining the dignity of Agricultural labor, enlightening its application and embellishing its abodes, we shall give new importance and stability to the arts of peace. If there were no other reason, Agriculture ought to rank first in the order of human occupations because its genius is pacific. Nothing could be more disastrous to us than a spirit of aggression or aggrandizement taking possession of the minds of our people. I know there is much in the extension of empire, which appeals to the strongest impulses of our nature —to the ambition of some, the cupidity of others, and the national pride of all. But if we may trust to the teachings of history, there is nothing in wide-spread dominion, which promises either lasting security or strength. The most extended empires are those which have fallen most suddenly and hopelessly into poverty and impotence. Gentlemen, there is a nobler national pride—one more worthy of our origin and our destiny—in elevating our internal condition, and devoloping our resources, in those improved applications of the mechanic arts which have just given us two distinguished triumphs in the face of the assembled world, in presenting to the other nations of the earth, an example of order and high civilization at home, and in maintaining in our intercourse with them a sacred regard for all the dictates of truth, honor, and international duty.

www.ingramcontent.com/pod-product-compliance
Lightning Source LLC
LaVergne TN
LVHW011147110826
845150LV00008B/2560

* 9 7 8 1 4 1 8 1 9 1 7 1 9 *